Let's Take a Hike!

Converting Fractions to Decimals

New Hanover County Public Library
201 Chestnut Street
Wilmington, NC 28401

Holly Cefrey

PowerMath™

The Rosen Publishing Group's
PowerKids Press™
New York

Published in 2004 by The Rosen Publishing Group, Inc.
29 East 21st Street, New York, NY 10010

Book Design: Ron A. Churley

Photo Credits: Cover © Ulf Sjostedt/Taxi; p. 4 © David Carriere/Index Stock; p. 6 by Michael Tsanis; p. 8 ©
Brian Payne/Index Stock; pp. 10–11 © Corbis; p. 12 © James Lemass/Index Stock; p. 12 (top inset) ©
Lauree Feldman/Index Stock; p. 12 (bottom inset) © Robert Franz/Index Stock; p. 14 © Inga Spence/Index
Stock; p. 14 (inset) © Ken McGraw/Index Stock; p. 16 © Bill Bridge/Index Stock; p. 18 © Diaphor
Agency/Index Stock; p. 18 (top inset) © Lewis Kemper/Index Stock; p. 18 (bottom inset) © Tom Walker/
Index Stock; p. 20 © Tom Stewart/Corbis; p. 20 (inset) © Photodisc.

Library of Congress Cataloging-in-Publication Data

Cefrey, Holly.
  Let's take a hike! : converting fractions to decimals / Holly Cefrey.
       p. cm. — (PowerMath)
Includes index.
Summary: Introduces different types of hikes, describes what you can see
on a hike, and explains how to convert fractions of walking distances to decimals.
  ISBN 0-8239-8979-8 (lib. bdg.)
  ISBN 0-8239-8928-3 (pbk.)
  6-pack ISBN: 0-8239-7456-1
  1. Fractions—Juvenile literature. 2. Decimal fractions—Juvenile
literature. [1. Fractions. 2. Decimal fractions. 3. Hiking.] I. Title.
II. Series.
  QA117 .C44 2004
  372.7—dc21
  513.26   2003001258

Manufactured in the United States of America

Contents

It is important to exercise to keep our bodies healthy and strong. Hiking is good exercise because it gets oxygen deep into our lungs and gets our hearts pumping.

Take a Hike!

Today, we are going for a hike. A hike is a walk in the **wilderness**. Hiking is great exercise, but we can also do and see many things on a hike. We can listen to chirping birds and look at beautiful plants. We can even catch grasshoppers and look at them up close.

Hiking is very popular in the United States. In fact, more than $\frac{1}{4}$ of all Americans go hiking each year! We can also write this **fraction** as a **decimal**. A decimal is just like a fraction, but it is written with a decimal point. To make $\frac{1}{4}$ into a decimal, we divide the top number (1) by the bottom number (4).

$$\begin{array}{r} 0.25 \\ 4\overline{)1.00} \\ -\,8 \\ \hline 20 \\ -20 \\ \hline 0 \end{array}$$

$\frac{1}{4}$ as a decimal is .25

More than .25 of all Americans go hiking each year!

Lewis and Clark Trail

The Lewis and Clark National Historic Trail begins in Illinois and goes through Missouri, Kansas, Iowa, Nebraska, South Dakota, North Dakota, Montana, Idaho, Oregon, and Washington.

A Famous Hike

Two **explorers** named Meriwether Lewis and William Clark made a famous hike that lasted from 1804 to 1806! Their hike took them across the western part of America, which was unexplored wilderness at that time. The path they followed is now a trail called the Lewis and Clark National Historic Trail.

The Lewis and Clark National Historic Trail is about 3,700 miles long and goes through 11 of the 50 states in the United States! We can write this as the fraction $\frac{11}{50}$. We can divide 11 by 50 to write this fraction as a decimal.

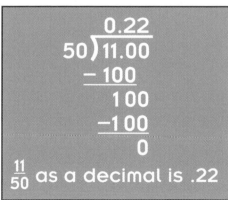

$$50\overline{)11.00} = 0.22$$

$\frac{11}{50}$ as a decimal is .22

The Lewis and Clark National Historic Trail goes through .22 of the 50 states in the United States.

It is good to be prepared when you take a hike. Some helpful things to bring on your hike are: a map of the area, a first-aid kit, a healthy snack, a bottle of water, sunscreen, bug spray, a jacket, a hat, and a flashlight.

Safety on the Trail

No one should hike alone. Hiking with our family and friends makes hiking safe, and it also makes it more fun. During our hike, we can talk with our hiking buddies about the different things we see along the trail.

It is important not to carry anything too heavy on a hike. We learned that it is best if the things a hiker carries weigh less than $\frac{1}{5}$ of their body weight. This is the limit of the extra weight that a hiker should carry. We can divide 1 by 5 to show this fraction as a decimal.

$$
\begin{array}{r}
0.2 \\
5\overline{)\,1.0} \\
\underline{-1\,0} \\
0
\end{array}
$$

$\frac{1}{5}$ as a decimal is .2

The Hike Begins

We're going on a hike in a park near where we live. The trail is 1 mile long and takes us through a forest and into a valley. The trail ends on top of a hill on the other side of the valley.

This garter snake is about $\frac{4}{5}$ the size of a full-grown garter snake. We can write this as a decimal by dividing 4 by 5.

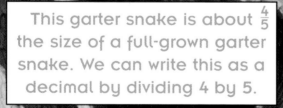

$$\begin{array}{r} 0.8 \\ 5\overline{)4.0} \\ \underline{-4\ 0} \\ 0 \end{array}$$

$\frac{4}{5}$ as a decimal is .8

Flowers and plants cover the ground where the trail starts. Animals such as the **garter snake** live underneath the flowers and plants. A garter snake can grow up to 30 inches long. It is not **poisonous**, but we know that it is a good idea to stay a safe distance away from any snake that we might see on our hike!

There are about 13 different kinds of garter snakes. At least 1 kind of garter snake can be found in most states in the United States.

prairie dog

chipmunk

ground squirrel

There are many different kinds of squirrels. Other squirrels that live in the ground are prairie dogs and chipmunks.

During our hike, we see grasshoppers hopping and butterflies flying. We also notice that there are small holes in the ground. We take a break to look for the animal that makes the holes. By now we have hiked $\frac{2}{5}$ of the trail. Since the trail is 1 mile long, this means we have hiked $\frac{2}{5}$ of a mile. We can divide 2 by 5 to show this fraction as a decimal.

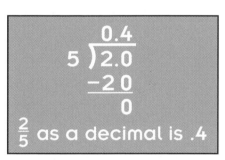

$\frac{2}{5}$ as a decimal is .4

We have hiked .4 mile so far.

Finally, we see a ground squirrel coming out of one of the holes!

Each kind of tree has its own special leaves. You can compare the leaves you find during your hike to see the differences between them. If you take a journal on your hike, you can draw or trace the leaves as well.

Trees and Leaves

We see grasses, plants, and small trees along the trail. Trees are very important to life on Earth. Their leaves help make the air clean for us to breathe. When we build towns and cities, we cut down trees and cause many changes in nature. Animals must find new places to live. Sometimes a type of animal dies out if it can't find new places to live.

There are thousands of different kinds of trees on Earth. Some **scientists** say that more than $\frac{1}{20}$ of them grow in the United States. We can divide 1 by 20 to write this fraction as a decimal.

$$
\begin{array}{r}
0.05 \\
20 \overline{)\,1.00} \\
-\ \ 0 \\
\hline
1\,00 \\
-1\,00 \\
\hline
0
\end{array}
$$

$\frac{1}{20}$ as a decimal is 0.05

There are about 9,700 different kinds of birds in the world!

Down in the Valley

When we get to the valley, our hiking group takes another break so we can all rest and look around. By now we have hiked $\frac{1}{2}$ of the trail, or $\frac{1}{2}$ of a mile. We can divide 1 by 2 to write this fraction as a decimal.

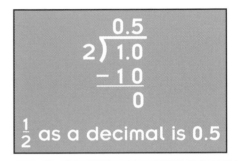

$$2\overline{)1.0}$$
$$0.5$$

$\frac{1}{2}$ as a decimal is 0.5

We have hiked .5 mile so far.

Many valleys have trees. Because a valley is lower than the land around it, rainwater and wind carry tree and plant seeds down into the valley from the surrounding hills. Birds also help to make plants grow by spreading plant seeds over the land. Birds eat plant seeds, but sometimes they drop seeds on the ground. Plants will eventually grow where a bird dropped some seeds.

white-tailed deer

turtle

There are about 35 different kinds of deer in North America and about 50 kinds of turtles in North America north of Mexico.

A clear stream runs through the valley. We come to a small bridge that is built over the stream and stop to look at the animals that live in the water. We see small fish in the water and a turtle sunning itself on a rock. Turtles are **reptiles**, which means they are cold-blooded. Their body stays about as warm or as cold as the surrounding air or water. We also see a deer in the distance.

By now we have hiked $\frac{3}{5}$ of the trail, or $\frac{3}{5}$ of a mile. We can divide 3 by 5 to write this fraction as a decimal.

$$
\begin{array}{r}
0.6 \\
5 \overline{)3.0} \\
-3\ 0 \\
\hline
0
\end{array}
$$

$\frac{3}{5}$ as a decimal is 0.6

We have hiked .6 mile so far.

arrowheads

Different Native American groups made different sizes and types of arrowheads.

We are almost finished with the hike. The trail is beginning to go uphill. By now we have hiked $\frac{3}{4}$ of the trail, or $\frac{3}{4}$ of a mile. We can divide 3 by 4 to write this fraction as a decimal.

We begin hiking up the trail and see some shiny black and gray rocks on the ground. In school, we learned that Native Americans once used these rocks to make **arrowheads**. Hundreds of years ago, Native Americans made arrowheads from stone or animal bone, then put the arrowheads on their arrows. They then used their arrows to hunt animals for food.

$$4\ )\overline{3.00} = 0.75$$

$$
\begin{array}{r}
0.75 \\
4\ )\overline{3.00} \\
-2\,8 \\
\hline
20 \\
-20 \\
\hline
0
\end{array}
$$

$\frac{3}{4}$ as a decimal is 0.75

We have hiked .75 mile so far.

The End of the Trail

The trail ends as we reach the top of the hill. We have hiked 1 full mile! We have seen many things on our hike and have learned about different animals, plants, trees, and rocks. Each hiking trip can teach us something new.

The National Park System has thousands of miles of trails that are great for hiking. Ask your parent or teacher to help you find out about trails near you. Then take a hike!

Common Fractions and Their Decimals	
$\frac{1}{10}$	0.1
$\frac{1}{5}$	0.2
$\frac{1}{4}$	0.25
$\frac{2}{5}$	0.4
$\frac{1}{2}\left(\frac{2}{4},\frac{3}{6},\frac{4}{8}\right)$	0.5
$\frac{3}{5}$	0.6
$\frac{3}{4}$	0.75
$\frac{4}{5}$	0.8

Glossary

rrowhead (AIR-oh-hed) The tip of an arrow. Arrowheads are usually made from stone or animal bone.

ecimal (DEH-suh-muhl) A fraction that is written with a decimal point.

xplorer (ek-SPLOOR-ur) Someone who travels to new places to learn about them.

raction (FRAK-shun) A smaller part of a whole thing. A part of a whole number.

arter snake (GAR-tuhr SNAYK) A small brown or green snake with yellow stripes. It is not poisonous.

oisonous (POY-zuhn-us) Containing poison. Poison is very dangerous to a person's health if it gets into their body.

eptile (REP-tile) One of a group of animals that is usually covered with scales and that can change how hot or cold its body is by moving into a warm or cool place.

scientist (SIE-uhn-tist) A person who studies the way things are and the way things act.

wilderness (WILL-duhr-nuhs) A place that has few or no people living in it and that has been untouched by humans.

Index

A
America(ns), 5, 7
arrowheads, 21

C
Clark, William, 7

D
deer, 19

E
Earth, 15
exercise, 5
explorers, 7

F
fish, 19
forest, 10

G
garter snake, 11
grasshoppers, 5, 13
ground squirrel, 13

H
hiking buddies, 9
hill(s), 10, 17, 22

L
Lewis and Clark National
    Historic Trail, 7
Lewis, Meriwether, 7

N
National Park System, 22
Native Americans, 21

R
reptiles, 19

S
stream, 19

T
trail(s), 7, 9, 10, 11, 13,
    15, 17, 19, 21, 22
trees, 15, 17, 22
turtle(s), 19

U
United States, 5, 7, 15

V
valley(s), 10, 17, 19

W
wilderness, 5, 7

ML   6/04